Carlos Eduardo Vidal da Silva
Ricardo de F. Cabral

Tetra Pak recycling for the construction industry

Carlos Eduardo Vidal da Silva
Ricardo de F. Cabral

Tetra Pak recycling for the construction industry

Use of aseptic carton packaging waste in non-structural concrete beams

ScienciaScripts

Imprint

Cover image: www.ingimage.com

This book is a translation from the original published under ISBN 978-613-9-63547-4.

Publisher:
Sciencia Scripts
is a trademark of
Dodo Books Indian Ocean Ltd. and OmniScriptum S.R.L publishing group

120 High Road, East Finchley, London, N2 9ED, United Kingdom
Str. Armeneasca 28/1, office 1, Chisinau MD-2012, Republic of Moldova, Europe
Printed at: see last page
ISBN: 978-620-7-67673-6

I dedicate this work to God, my family and my son Eduardo Vidal.

ACKNOWLEDGEMENTS

I would firstly like to thank God, the creator of heaven and earth, for opening yet another door that no man can close and for shaping this achievement in my life.

To Prof Dr Ricardo de Freitas Cabral, for his guidance, teachings, encouragement, dedication and patience.

To Prof Dr Roberto de Oliveira Magnago, for his friendship, comments and constructive suggestions on this work.

To Prof Dr Horácio Guimaraes Delgado Jr and all the other professors for their knowledge and suggestions for this book.

To Prof Carlos Eduardo Abreu de Lima de Souza, for his friendship, words of encouragement and advice, which added a lot to the completion of this work.

To the technicians of the Engineering and Materials Sciences laboratories of Arthur Almeida Bittencourt and Dirceu Hartung de Camargo Coutinho, who played a major role in this book. In addition to carrying out the tests.

To Ana Maria, for her generosity, dedication and attention, who didn't let me give up on my master's degree, and to everyone else at the MeMat secretariat and UniFOA librarians.

To my fellow Master's students, who contributed with their experiences of their own work.

To my beloved family, my father Carlos Augusto Vidal, my mother Inaura Martins, my brother Rodrigo Vidal.

In particular, my girlfriend Hana Caroline and my son Eduardo Vidal, for their affection, care, loyalty, love and for building my life. Thank you from the bottom of my heart! I love you!

To all the people who contributed to this project, thank you very much!

SUMMARY

The practice of using recyclable materials in construction is becoming more common every day. This positive trend aims to build a sustainable legacy for new generations, in line with the new world order. According to CONAMA resolution 307, all municipalities must provide environmentally appropriate management of construction waste. Therefore, the aim of this work is to analyse the confluence of mechanical properties in concrete, specifically the use of cardboard packaging fibre, by subjecting the fibre to the axial compressive strength test, according to NBR12118, over a period of 7, 14, 21 and 28 days. From natural concrete to the use of cartonboard strips cut with a guillotine to form strips with a thickness of 30mm x 3mm. This work is moving towards the great challenge of adding carton packaging to natural aggregate in the manufacture of concrete for the final construction product. The results of the 28-day compressive strength test were 24.5 MPa for 100% natural aggregate, 22.53 MPa for 5% carton *packaging,* 20.09 MPa for 10% carton packaging and 16.22 MPa for 15% strips.

Keywords: Composite, Aggregate, Cement. Concrete, Strips.

SUMMARY

CHAPTER 1

INTRODUCTION

The aim of this book is to develop a modular prefabricated construction system made from cylindrical specimens, sand and gravel, using cardboard packaging boxes as the raw material. The proposal to develop this construction system allows it to be easily inserted into joint efforts to make environmentally friendly buildings feasible, since the concrete beams can be produced on a large scale using renewable inputs.

The aim is to study the influence of adding carton packaging to concrete without replacing any of its components. In this way, it will be possible to propose a new material with different characteristics to conventional ones, with the aim of making concrete that is more economically viable for civil construction, and a considerable reduction in the final cost of the work, as it is a recyclable and easily accessible material, as well as unburdening the environment and adding a material that is easily accessible in Brazilian homes and industries, with the possibility of adding it to concrete in the 1:2:3, adding percentage dosages of carton packaging strips with the percentages in dosage traces (5%, 10%, 15%), of substitution with *carton packaging* strips, conforming test specimens of the final product, meeting the standards of ABNT, Brazilian Association of Technical Standards (NBR 5738 and NM67). The results obtained in the study will show whether the material will have mechanical resistance within the parameters of civil engineering and materials science.

CHAPTER 2

OBJECTIVE

The aim of this work is to develop and produce concrete mixes with the addition of aseptic carton packs and to evaluate them using the following tests:

- Mechanical resistance to compression;
- Water absorption content;
- Scanning Electron Microscopy (SEM) analysis;
- Analysis of the results obtained;
- Emptiness index

CHAPTER 3

BACKGROUND

The disposal of carton packs, as well as other solid waste, was one of the issues debated at the United Nations conference on the environment and climate. According to population growth, a key factor emphasised in debates on sustainability and the environment, the greatest environmental impacts are not caused by the population, but by consumption patterns. Paradigms in the industrial production line generate waste that causes environmental degradation, which calls for the correct disposal of waste. (UEMURA, 2017)

Pereira (2008) defines that affirmative actions aimed at preserving the environment, which were carried out in isolation in the past, are now more systemic and frequent. Legislation together with social appeal has led industries to seek standards through clean technologies, environmental management systems, social responsibility and reverse logistics.

The cost-benefit ratio is indispensable for the potential investment of private initiative and public authorities in the construction of sustainable projects, making them part of the process of re-educating the population itself, leaving a relevant legacy for generations to come. However, developing a profitable partnership for the management of cooperatives, groups of entrepreneurs, builders, with the bias incorporated into low-cost social housing programmes, without losing the comfort of the product marketed to the consumer.

In accordance with Law 12.305/10 on the National Solid Waste Policy (PNRS), which provides for a reduction in waste generation through sustainable practices in domestic and industrial consumption, with mechanisms that make it possible to increase the reuse of solid waste.

CHAPTER 4

LITERATURE REVIEW

Based on NBR 5732, CPIII-40 Portland Cement is a hydraulic binder obtained from a homogeneous mixture of portland clinker and granulated blast furnace slag. Indispensable in the construction industry, cement is used in urban construction work, low-income housing, hospitals and large developments such as hotels and commercial buildings, among others. Concrete, whose formulation includes cement when dissolved with sand and gravel, is characterised by its mechanical strength, which comes from the compression and rupture of specimens that are applied with mortar (BAUER, 2011).

For Castro (2011), the chemical nature of the additive in concrete influences the preparation of mortar-based materials in the fresh state. In current construction concepts, additives such as aseptic carton packaging, PET and other additives, when incorporated into the mix, interconnect with the various components of the cement and influence its hydration reactions. In order to apply these additives satisfactorily, it is necessary to know the basics of cement chemistry and the forms of cement-additive interactions.

4.1 Cement

4.1.1 History and Evolution

According to Ambrozewicz (2012), the meaning of cement in Latin is *caementu,* which translates into stone from rocks. In Ancient Egypt, calcined plaster was used for monumental sculptures. However, the mass obtained differed from today's mass in that it contained the addition of limestone and gypsum, resulting from the decarbonisation of the stone with the intervention of fire.

According to Callister (2016), the Greco-Roman civilisation kept the true formula of its cement secret for generations. With the fall of these ancient peoples, the secret was lost in time, which had the detrimental effect of damaging the properties of the cement's components. As a result of this loss, the material had to be reformulated.

John Smeaton developed a low-temperature calcined cement capable of resisting seawater, commissioned by the English government in 1756, which would later be called portland cement (Coutinho, 2016).

According to Betsuyaku (2015), concrete with cement in its formulation, when mixed with small aggregates, increases its compressive strength, making it possible to mould structural parts.

Coating and laying mortars also use cement as their fundamental raw material. Cement has played a major role in the development of humanity, especially in the construction of large urban centres.

4.1.2 Aggregates commonly used for concrete

Aggregates are components of the construction industry that make up the mixture of cement and water used to form concrete. They are classified as coarse and fine: coarse aggregates are defined as grains that easily pass through a square mesh sieve with a nominal opening of 152mm and remain contained in the sieve suggested by the ABNT 4.8mm. In the case of fine aggregates, sand of natural origin or derived from the crushing of stable or mixed rocks and its grains pass through the ABNT 4.8mm sieve and remain retained on the ABNT 0.075mm sieve, in accordance with the current Brazilian regulatory standard (ABNT NBR 7211).

According to Cunha (2015), the use of aggregates has become increasingly popular over time. Ancient civilisations such as the Egyptians, Mayans and Chinese already used aggregates. The evolution of new techniques and the discovery of new technologies contributed to the development of diverse applications, including the most famous of all, concrete.

In agreement with Neville (2016), classification and characterisation has enabled humanity to make great progress, as it has led to the discovery of new products in civil construction, presented by different models of concrete and other mixtures, for soil stabilisation, for paving, in addition to its ornamental use, which has a large economic portion in its use.

It is classified as synthetic or natural. Synthetic is classified as sand and stones made from crushed rock. However, it requires human intervention to change the shape of its particles. On the other hand, natural sand comes from rivers or river ravines and stones (SILVA, 2015).

However, the ratio of aggregates is characterised by the mass present, such as fine and common. Fine aggregate is made up of expanded clay, pumice, vermiculite and common aggregate takes the form of crushed stone, sand and pebbles, full-bodied aggregates such as hematite, magnetite, barite.

Aggregates mixed with other materials are highly relevant in various tests obtained for the defined application, such as granulometry, peculiar mass, fineness modulus, clay at high temperatures, organic impurities and powdery materials. They are fundamental in the manufacture of mortar and concrete (LACERDA, 2015).

4.1.3 Types of cement

According to Ambrozewicz (2012), the definition of cement is represented by the acronym CP (Portland cement), accompanied by numbers from I to V that classify the type of cement; in addition to the classes represented by the numerals 25, 32 and 40, which indicate the compressive strength of the standardised test specimen.

Cement agglomerates made up of clinker, a product resulting from the calcination of limestone and clay and its chemical components, are distinguished by the addition of gypsum when the aim is to improve the cement's setting time. Another important component in the structure of cement is slag, whose purpose is to increase the degree of durability, but the presence of sulphate in high proportions minimises the strength of concrete, while pozzolanic clay has the potential to waterproof concrete. Another fundamental component in the composition of cement is limestone, the incorporation of which is intended to lower the economic cost without reducing the quality of the material (TORRES, 2017).

It should be emphasised that the components of cement vary according to the region and the characteristics of local factories and industries. The distinction between the elementary types of cement is illustrated in Table 1.

Table 1 - Composition of Portland Cements

COMPOSITION OF PORTLAND CEMENTS Composition (% by mass)						
Type of cement	Classification	Clinker + Gypsum	Blast furnace slag(E)	Pozzolanium slag (Z)	Carbon material (F)	Norma
Common	CPI CPI-S CPII-E	100 99-95 94-56	- 6-34	-	- 0-10	NBR 5732
Compound	CPII-Z CPII-F	94-76 94-90	- -	6-14 -	6-10	NBR 11578
Blast Furnace	CPIII-40	65-25	35-70		0-5	NBR 5735
Pozzolanic	CPIV	85-45		15-50	0-5	NBR 5736
High initial resistance	CPV-ARI	100-95	-	-	0-5	NBR 5733

Source: Ambrozewicz, 2012, p.84.

According to table 1, cements can be classified as:

- **CP I** (Ordinary Portland Cement) is manufactured only with the addition of gypsum, the component responsible for the loss of fluidity of the paste called setting, based on NBR5732.
- **CP II** (Composite Portland Cement) Resulting from the analysis of ordinary Portland cement, a format was developed with added slag or pozzolan components, making it a competitive product that accounts for 70% of the industrial market in Brazil. It is used on a large scale in common applications, directly replacing CP I based on NBR 11578.
- **CP III-40** (Blast Furnace Portland Cement) The result of the addition of steel slag, the fusion of metals processed in the blast furnace with the appearance of silicate. Steel slag plays the role of representing the latent hydraulic characteristics, but the hydration activities of the aforementioned product are slow. This material is often used in the construction and paving of airport runways, maritime works and environments, dams, reservoirs and sewage plants based on NBR5735.
- **CP IV** (Pozzolanic Portland Cement) Made by combining steel furnace slag and pozzolan. Unlike slag, pozzolan does not react with water. During the grinding process, it reacts with the calcium hydroxide component associated with the hydraulic binder at room temperature, resulting in compounds with mass properties based on NBR5736.
- **CP V- ARI** (Portland Cement of High Initial Resistance) Its main characteristic is that it achieves high immediate resistance after application. This is due to the possibility of using a specific amount of clay and limestone in the manufacture of the clinker, the refinement in the grinding for the cement when reacting with water acquires an increase in the rate of resistance and speed based on NBR5733.

4.2 Concrete with fibres

4.2.1 Fibres

Concrete is a structural material used worldwide in the construction industry. It has gained prominence and provoked discussion and study on the subject, and is widely debated among researchers. It offers satisfactory costs and benefits, high durability, resistance to compression and high temperatures, pre-moulding, architectural versatility and acoustic control. However, concrete has a number of limitations due to its fragility and low efficiency in deforming the composite before breaking. Although its tensile strength is lower than its compressive strength, working together with steel reinforcement provides the structural foundations with the necessary tensile strength and ductility (BAUER, 2011).

In order to reduce these limitations, the addition of reinforcing strips to cement mixtures has made progress in recent years. Unlike conventional reinforcement, which is localised and requires prior assembly, the strips are mixed directly into the concrete and distributed randomly, optimising the time, labour and cost of certain applications compared to the traditional process. The addition of strips to cement mixtures provides remarkable progress in various mechanical properties, emphasising toughness, resistance to fatigue and impact. Their main characteristic is to accelerate energy absorption capacity, as they act as a bridge to carry tensions across cracks, slowing down their propagation and expansion. Among other functions, concrete reinforced with strips has greater ductility compared to non-reinforced matrices, which become ineffective after the first crack has formed (SALVADOR, 2013).

Various types of fibres are used to reinforce concrete, the main ones being steel strips, polymeric strips and natural strips. Several recent studies on the importance of steel strips in concrete behaviour have been published (FATHI, 2014).

The influence of polymeric fibres on the performance of concrete has been analysed by various researchers from different aspects. The addition of polypropylene strips significantly improves the mechanical properties of concrete, including tensile strength, toughness and abrasion resistance. The compressive strength with the addition of polymer strips shows a reduction in variation (KHALOO, 2014).

4.2. 2Basic aspects

The fibres in concrete act on the cracks as the cement paste matures, protecting the appearance of macro-cracks, as well as inserting themselves into the stiffened paste, working as a barrier to the development of crack entry and length. Multiple events occur in concrete with strips, the most important of which are derived from the concrete matrix to physical and geometric factors, the content of the strips used and the composition between the strips and the matrix. Pouring and densification processes are important phenomena, as they directly affect the distribution and direction of the strips in the matrix.

New research and analyses of concrete with strips have emerged. The choice of this composite is higher all over the world and scientists have made various types of strips available on the market: steel fibre (straight, wavy, twisted, deformed at the poles with hooks), polymeric strips (PP, polyester, nylon and aramid), glass strips.

According to Nosheen (2018), the application of fibres in concrete began in 1911, when Grahan analysed the addition of steel strips with the idea of increasing the strength of reinforced concrete. In the 1960s, technical and scientific improvements were made and various applications and practices of fibrous concrete appeared on the market.

Metha (2010) emphasises the emergence of the first structural concrete, created in 1971 to produce pre-moulded panels measuring 3250mm^2 and 65mm thick. The concrete contained 3% cold-drawn steel fibre, 0.25 mm in diameter and 25 mm long. The panels were used in the car park garage at London Heathrow Airport. In this situation, concrete with steel strips has become very popular in the following applications: industrial floors, pavements, tunnel linings, anchoring blocks for protection cables and other areas where stresses condense, rainwater pipes, sewers, culverts, shells, roof tiles, retaining elements, sheet piles, elements and structures subjected to earthquakes, elements subjected to high impact, sleepers, manufactured structural elements in general.

According to Hwang (2016), the strips should be applied with full use of the elements, where the exchange of stresses is very relative (pavement and tunnel lining) covering all three dimensions; highly decorated elements where the extremes between the reinforcements make concreting impossible; materials of smaller proportion where the inadequate positioning of conventional reinforcement can unintentionally alter the appropriate height, in elements with little development.

4.2.3 Concrete matrix

Strip concrete is made up of hydraulic cement, water, fine aggregates, coarse aggregates and discrete discontinuous strips, often with chemical additives and mineral additions to enrich the strength and workability processes.

According to Carasek (2007), the type of cement with the addition of strips to the concrete is very well accepted. However, this cement must comply with the intended use and resistance. However, the large-scale dimension is significant to the action of concrete with strips, as each stage of this concrete should not exceed more than 20mm and not more than 10mm so as not to alter the uniform organisation of the fibres. There is a risk of deceleration between the strips and the alkalis in the cement.Figueiredo (2013) points to the proportions of the aggregate, the higher the length of the aggregate, the greater the problems of intervention in the concrete, positively interfering with the use of the fibre. It is recommended that there be affinity in the space between the aggregates and the fibres, so that the fibres capture the cracks that occur in the interstice of the composite more frequently. This event is shown in figure 1 below:

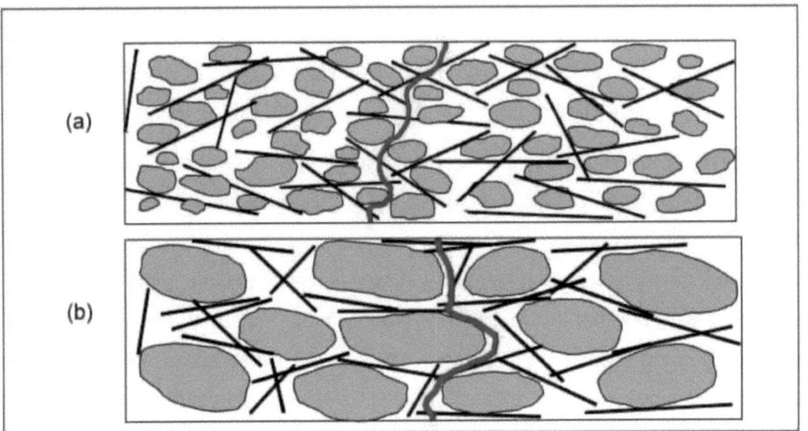

Figure 1 - Concrete with strips where there is (a) and where there is not (b) dimensional compatibility between the strips and the coarse aggregate
Source: Figueiredo, 2013.

Illustration of the action of the fibre as protection for the concrete and not just as an aid to the industrialised mass. Therefore, this compatibility is of great importance, as cracks emerge specifically in the interface region as a result of the coarse aggregate and the concrete of lower and moderate mechanical resistance. The length of the fibre can be twice the size of the coarse aggregate, suggesting approximately 2.5 to 3 times the length for the fibres to act through the stress transmission link in the cracks (CALLISTER, 2016).

4.2.4Composite and fibre-matrix interaction

The use of additional water reducers is common in strip concrete. This application of mineral additives, such as microsilica, has been commonly used in these concretes. The use of microsilica favours a potentially dense matrix, improving the fibre-matrix process and the mechanical characteristics of the concrete (CASTRO, 2011).

It is important to highlight the structural material aspect, which has a meticulous balance in reducing fibre-matrix adhesion. In this way, the strips have a repressed bond with the matrix and can pass under low loads, contributing ineffectively to reducing cracking. As a result, the strips reduce the toughness of the system. Consequently, the strips behave like inefficient joints, developing a superficial advance in mechanical properties.

The fibre-matrix bond is subject to various factors, such as: fibre-matrix friction, mechanical gearing of the matrix to the fibre, and the physical-chemical junction between the materials. These phenomena are triggered by the peculiarities of the strips (volume, modulus of elasticity, strength, geometry and orientation) and start solely from the matrix (composition, cracking episode and physical-mechanical properties).

Before the matrix cracks, the dominant process is the transition of elastic stresses and the distant path of the fibre and matrix in the modality, with the interaction between the systems being geometrically compatible. In the more advanced stages of loading (tensile or bending phenomena), microcracks appear and the stresses are quickly stored at the ends of these cracks, causing the process to accelerate and the input to rise, ending up in the sensitive sharing of the material.

Tensile fragmentation of fibre concrete occurs when the strips expand plastically or elastically due to the concrete matrix slumping in the fibre-matrix transfer zone, when the fibre stagnates or when it breaks.

The strength of fibre concrete is a classification of the profile required by the direction of the strips, and is not standardised. However, with vibrated concrete, the strips act with priority influence perpendicular to the concreting location. The compilation refers to differentiated counselling, highlighting the use of surface vibration (direction parallel to the form), but this event is carried out on site. The internal vibrator receiver must have an abundance of paste and a low fibre content (HWANG, 2016).

Schettino (2015) mentions the interference factors of the fibres, as well as the geometric characteristics, the volumetric quantity embedded in the concrete, the mechanical resistance in manufacture, the tension and adhesion in the interstice of the fibrous matrix and the ratio between the diameters of the fibre and the aggregate. It should be emphasised that the primary elements of concrete are altered by the addition of strips.

4.2.5 Workability

The lack of functionality of strip concrete relates to the volumetric storage of strips. However, the shape of the strips, the characteristics of the machine, the mix selected in the manufacture of the melt, the profile and the quantitative form of plasticiser used in the mix, in order to affect the workability of the concrete (CALLISTER, 2016).

The addition of strips affects the solidification factors of the composite and its workability. As a result of the addition of the strips to the concrete, a large surface area is added that requires wetting water, which means that the larger the shape of the strips, the greater the ramming of the concrete (CUNHA, 2015).

Metha (2010) points out that the decrease in texture of concrete with fibres, projection and compaction are more satisfactory than in conventional concrete with low consistency strips. Three methods are highlighted to prove the applicability of the composite with strips in the fresh state and they are as follows:

- Cone trunk subsidence;
- Inverted cone trunk;
- Vebe test, in which the variation in the consistency of the concrete is labelled as the appropriate moment to remould the concrete contained in the equipment from a truncated shape to a cylindrical shape, as shown in figure 2 below:

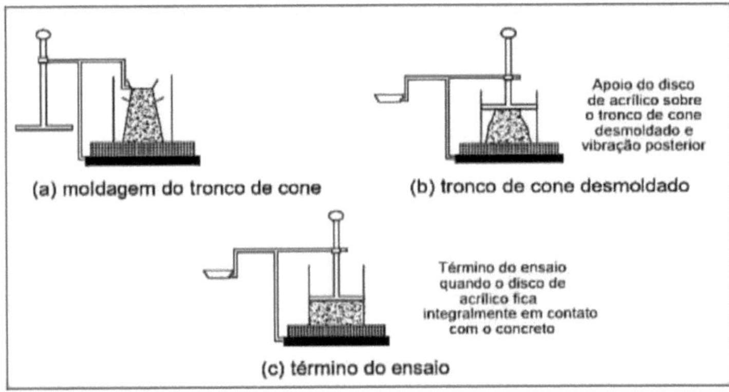

Figure 2 - Vebe test
Source: Metha, 2010.

4.2.6The effect of fibre geometry

There are a variety of manufacturing processes for steel strips, the most common of which is profiled wire cut from steel with a lower carbon content. Most steel strips are made from carbon steel, but those containing metal alloys are more resistant to corrosion, but are more efficient for use in refractory concrete and nautical structures.

The geometries of steel strips are extremely varied. Flattened steel fibre has parameters ranging from 0.15mm to 0.64mm thick and widths ranging from 0.25mm to 2.0mm. This geometric standardisation of the fibre consists of the resultant between length and equivalent diameters, normally presenting values in the range of 20 to 100. When shorter, the length of the fibre reduces the cross-section and the thickness is of little value in the process. The thicker the fibre, the higher the resistance capacity after the concrete cracks (RIBEIRO, 2018).

In the case of corrugated steel strips, these are available both in length and at the ends. However, the steel strips are glued together with specially water-soluble adhesives, producing bundles of 10 to 30 strips to establish the optimum degree of handling when adding the concrete (BENTUR, 2007).

In the view of Castanheira (2017), cement composite materials such as aggregates are naturally selected for the insertion of polymeric materials and, in addition, they offer good cost-benefit ratios, while demonstrating difficulties in terms of ductility, resistance to rupture and capacity to absorb deformation energy.

The polymeric strips are divided geometrically into micro-strips and macro-strips with the addition of polypropylene micro-strips (micrometre equivalent diameter and thickness close to unity). In Brazil, there is a common preference for optimising cracking by retraction and controlling exudation, although the strips do not have structural characteristics, while the polypropylene macro-strips (millimetre equivalent diameter and structure varying between 20 and 100) are called structural strips and compete with steel strips. In North America and Europe, these strips are widely commercialised on the market, unlike in Brazil (MOREIRA, 2017).

Patil (2015) argues that the reduction in cracking and exudation with the application of polymer strips is due to the fact that the strips disrupt the hydraulic flow in the concrete core, increasing its cohesion. This change can lead to a number of specific applications such as the engineered composite, avoiding the risk of spalling (an aggressive pathology in concrete due to chemical attack and design flaws) and proving the dimensional stability of the newly formed concrete.

Steel fibres are more common and effective for concrete, while polymer strips are more commonly used for specific jobs, such as architectural or decorative concrete, which has the characteristic of reducing the visual impact of the strips. In this adverse perspective, polypropylene and polyester or nylon strips are more commonly used.

There is a discrepant superiority of steel fibre when compared to polymer fibre. However, polymer strips are guaranteed to offer considerable cost-effectiveness for the construction industry. However, steel fibre offers greater performance and efficiency at a significantly higher cost (METHA, 2010).

4.2.7 The effect of fibre strength

The main characteristic of applying strips to concrete is that they do not change its compressive strength. However, the performance of the strips is to transmit the stresses through the cracks, as they come from shear, as occurs in the compression test. Concrete with strips shows that the volume of strips produced is less than 2%, which means that the properties of compressive strength, modulus of elasticity and maximum stress show a slight increase when subjected to traction and bending. The increase in the volume of the strips leads to both an increase and a reduction in the strength and modulus of elasticity. The volumetric reduction of the strips is detected in terms of the negative factor, such as high air content, influenced by the addition of strips in the matrix being preponderant (WU, 2018).

While the compaction of the matrix related to the storage of the dry aggregate mixture and the use of the appropriate mixer and vibrations, the increase in strength and modulus can be seen in the performance of strips in high quantities, providing the increase in compressive strength with the presence of steel fibres is 25% and for volumetric levels of fibre is 2.0% (HWANG, 2018).

According to Alves (2013), compression experiments on cylindrical specimens were carried out at three different dosages, with the addition of steel fibre, using hooks at the ends, with a length of 30mm, diameter of 0.62mm and structure 48, at 0%, 0.75% and 1.50% volumetric levels. The results indicate that the addition of fibres does not result in an increase in compressive strength, with the change not exceeding 16%.

High-strength concretes with added fibres are differentiated by the increasing axis of the compressive stress versus specific deformation curve, namely: strength, modulus of elasticity, deformation relative to maximum stress. They are therefore related to normal strength concrete. However, for both low-strength and high-strength concrete, the post-peak result is different from that of concrete without strips, showing an increase in ductility, as can be seen in the comparison between the stress curve and specific deformation, shown in Figures 3 and 4 below:

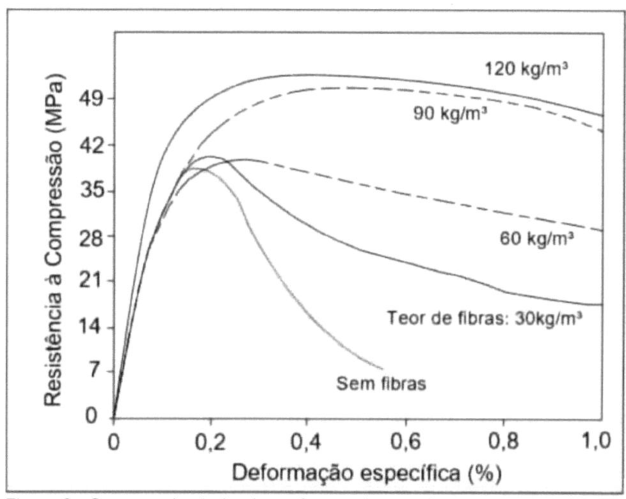

Figure 3 - Compressive behaviour of normal strength concrete with steel fibre
Source: Metha, 2010.

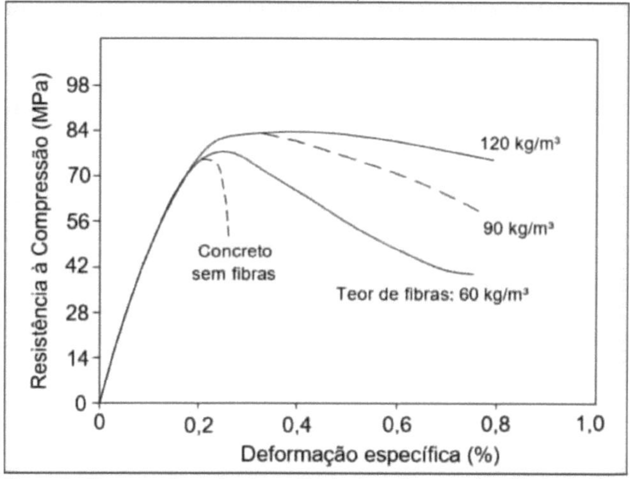

Figure 4 - Compressive behaviour of high-strength concrete with steel strips
Source: Metha, 2010.

The post-cracking energy effort due to compression of the matrix will, in turn, have different functions depending on the location of the fibres. Compacting the concrete perpendicular to the location of the strips will result in an increase in the post-cracking energy force, while concrete compacted perpendicular to the location of the fibres will have a greater post-cracking energy force than concrete compacted parallel to the desirable location of the strips (WU, 2018).

The tensile strength of strips in concrete is obtained through three different tests: the direct tensile test, the indirect tensile test, which consists of the diametrical compression tensile test, known as the Brazilian test, and the flexural tensile test.

The most important experiment for calculating the tensile strength of concrete is the direct tensile test. However, this test requires the application of high-precision adhesives and is the most difficult of all the tests to carry out, which is why it is only used in scientific papers. Although there is no defined rule on the best or worst test for the tensile strength of concrete, it is defined as a standard in the calculation processes of concrete structures, exemplified by NBR 6118 (2007) for accounting for cracking and minimum reinforcement of the strength and rupture of materials with the absence of transverse reinforcement and joint tension, this event being judged from sentences that classify with the compressive strength.

In the context of direct tensile strength, there is no defined methodology for the direct tensile test, resulting in different varieties of specimen moulds and support conditioning (NAAMAN, 2003).

According to Bentur (2007), once the effects of the strips used in the practical exercise are included (1% by volume), direct attraction is limited to 20% when multiplying resistance and excesses are quantified by using thicker strips.

According to Dias (2018), adding 1.5 per cent of strips to the cement-based matrix results in a 30 to 40 per cent increase in direct tensile strength.

Matrices with strip groupings (high-strength concrete with the inclusion of fly ash, for example) allow for an additional increase in tensile strength test results.

The strips lined up and located in the tensile stresses produce satisfactory increases in direct tensile strength, surpassing the strips that are occasionally distributed in the concrete matrix.

4.3 **Composition of *carton packs***

Carton packaging, which is one of the derivatives of the reference name, the largest manufacturer of "long life" packaging, this product began in 1952 in Sweden with the creation of a parallelepiped-shaped package made up of layers of paper and plastic, initiated by Ruben Rausing, with the aim of storing perishable liquid food (Fiorelli, 2008).

In 1961, the packaging as it is known today was created, with layers of paper, plastic and aluminium. In Brazil, however, the commercialisation of this product began in the early 1970s and was very popular with both producers and the consumer market, as the packaging allowed for the perfect preservation of perishable foods, such as milk, juice storage, which was very helpful due to the ease of preserving quality during transport.

Aseptic carton packaging, also known as long life, tetra pak or multilayer, is made up of six layers of different types, with three types of material, 75% paper, 20% plastic and 5% aluminium. The distribution of the following materials is divided into three layers of materials: polyethylene, paper and aluminium (Ajam, 2013). The protective layers are shown in Figure 5.

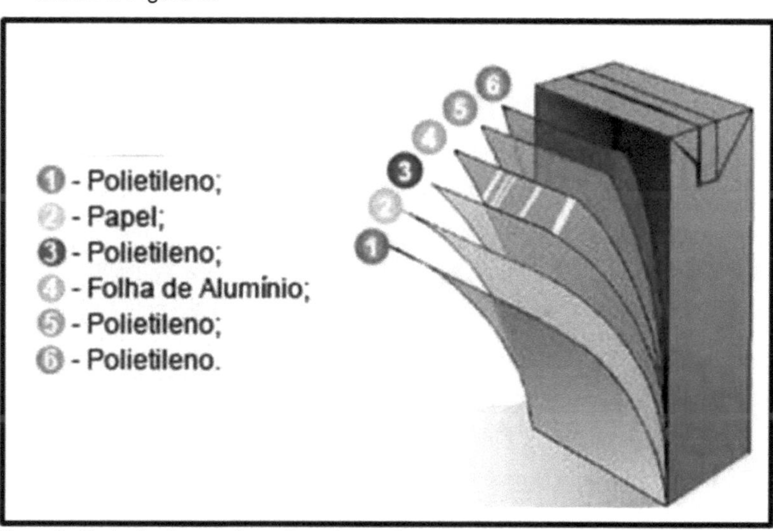

Figure 5 - Protective layers of aseptic carton packs

Source: (AJAM, 2013)

According to Cempre (2015), each material in this structure has a specific function:

1. Polythene layer: This protects the packaging from external moisture.

2. Layers of Duplex Paper: The white paper is used to allow printing and the brown paper ensures stability and rigidity, guaranteeing the structure of the packaging. It represents 75% of the entire structure of the packaging material.

3. Polyethylene layer: to provide adhesion between the paper and aluminium layers.

4. Aluminium layer: If this layer didn't exist, it wouldn't be aseptic. Its function is to prevent

air and light from entering the packaging, loss of flavour and contamination.

5. Polyethylene layer: to provide adhesion between the polyethylene and aluminium layers.

6. Polyethylene layer: This has a protective function, preventing the food from coming into contact with the aluminium.

According to López (2016), the plastic is called low-density polyethylene and is a partially crystalline polymer. It is distributed between 50 and 60 per cent in its production process, with pressure varying between 1,000 and 3,000 atmospheres and temperatures reaching 100 and 300 °C.

The characteristic material called Low Density Polyethylene (LDPE) combines a single grip characterised by high flexibility, high impact resistance and good processability. Aluminium stands out for its abundance and structural uses. There is a consensus that aluminium is in abundance and is used in our daily lives for cosmetic packaging, hospital supplies, as well as in the automotive industry and construction (Pedroso, 2007).

Paper is one of the most widely used items in everyday life. It is extracted from the wood pulp of trees such as pine and eucalyptus. Paper is made from fibrous elements of plant origin and is characterised by its high cellulose content, low cost and easy accessibility.

6.1.1 Recycling aseptic carton packs

The increase in population and industrial growth is characterised by an increase in the amount of organic and inorganic waste in society. With the amount of rubbish produced en masse, recycling is becoming increasingly important. According to the Business Commitment for Recycling (CEMPRE), Brazil has high rates compared to other Mercosur countries. Data shows that in 2015 Brazil was responsible for recycling 21% of *Treta Pak* packaging, totalling more than 59,000 tonnes. It should be noted that each tonne of recycled *Treta Pak is* around 680 *kilos* of *krafit* paper*, and* a high number of these packages are expected to be recycled, due to the selective collection initiatives organised by private initiatives, cooperatives and communities in a large part of society, providing an apparatus for the development of new technological processes (CEMPRE,

2016).

As well as recycling *Treta Pak cartons into Krafit paper, it* also produces components for corrugated cardboard, paperboard, egg cartons, shoe insoles, tissue paper and other components. Concerned about the environmental impact, another reuse technique is the manufacture of roof tiles, slabs, concrete beams and road paving using *carton* packaging because of the durability of the material, easy access to material collection and low cost for construction.

6.1.2 Adding carton packaging strips to concrete

Conceptually, the addition of *Treta Pak* strips to concrete is a relatively innovative compound. The early 1960s saw the emergence of new products derived from polymeric, metallic, vegetable and mineral strips. A line of research involving different types of strips began in the United States in 1971. The study was aimed at reinforcing construction materials (LÓPEZ, 2016).

The joining of fibrous composites in Brazil is even more recent on an industrial scale, having been widespread since 1990. Nowadays, the consumption of strips of various shapes and sizes has been widespread, produced in steel, plastic, glass and natural materials, although steel strips have been standing out in the industrial market, gaining positive acceptance in civil construction (CEMPRE, 2016).

According to Ajam (2013), strips are discontinuous elements with a length greater than the cross-sectional dimension and can be made from a variety of materials, from natural fibres such as sisal to steel, glass, polymers, pet and *Treta Pak.* The classification of strips can be characterised by size, varying in short or long parts, and by their presentation as loose, glued, monofilament and fibrillated.

The strips are classified in two different ways: short and long. Short strips act directly on the mortar, optimising a large part of the cracking in the composite over the load, without altering the resistance and workability to any significant degree, and can thus increase ductility levels. Long strips interfere with the concrete matrix by reducing macro-cracks, but reduce workability, giving the concrete a high tensile strength (METHA, 2010).

Nosheen (2018) highlights the importance of the cemented fibre-matrix in general concepts, highlighting two relevant effects: the first is characterised by toning the composite over all forms of transport by positively inducing tensile stresses. This means indirect traction, bending and shear. The second is a significant improvement in ductility and the cohesion of the fibre matrix with a tenuous characteristic.

The inclusion of strips has a major impact on flexural tensile strength. There are records of an increase of over 100 per cent in strength at high levels of fibre content. Being very efficient in the workability of concrete, the fibre goes beyond the barriers of the cracks that make up the concrete, interfering in the operation of polarised loads or when faced with climatic changes (SALVADOR, 2013).

The addition of *Treta Pak* packaging to concrete causes a reduction in its specific mass. Considering that *Treta Pak* is made up of light, less dense materials compared to sand, this results in a reduction in the specific mass.

CHAPTER 5

MATERIALS AND METHODS

5.1 **Materials used**

5.1.1 CPIII cement

Throughout this project, including the preliminary studies and the initial study, *standard CPIII-40 Portland* cement was used *(figure 6), which is* commonly used in construction work. Its main characteristic is the high content of steel slag, a mixture of metals resulting from the blast furnace process and the presence of silicate (Torres, 2017).

Composition of CPIII cement: (AMBROZEWICZ, 2012)

- Clinker and Gypsum: 25 to 65%
- Slag: 35 to 70%
- Limestone: 0 to 5%

Figure 6- CP III-40 cement
Source: (AUTHOR, 2017)

5.1.2 Brita

Gravel 1, a natural type with a size of 19mm, basically composed of granite and extracted from the Volta Redonda quarry, was granted by UniFOA.

5.1. 3Sand

The standardised medium sand purchased from the UNIFOA laboratory for making the test specimens has an average size of 2mm to 0.6mm, as shown in figure 7 below.

Figure 7 - Sand
Source: (AUTHOR, 2017)

5.1.4 Aseptic carton packs

The strips of carton packaging from the juice box collection were subjected to the guillotine cutting mechanism in strips measuring 30mm x 3mm, as shown in figure 8 below.

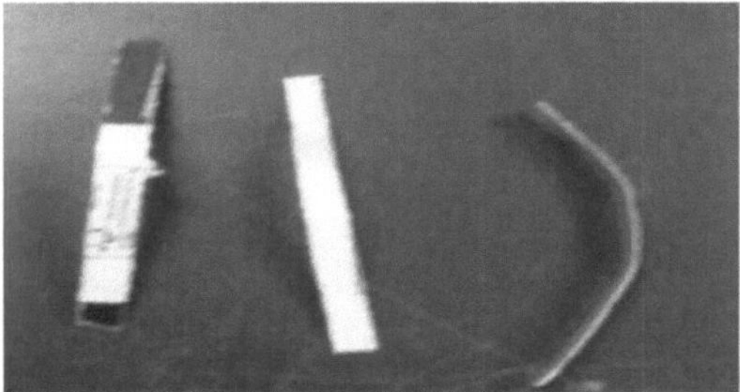

Figure 8 - Filament for aseptic carton packs
Source: (AUTHOR, 2017)

5. **2Methods**

This chapter sets out all the techniques implemented involving the process of characterising the microstructure of the material researched for this study. Figure 9 shows a flowchart that graphically displays the entire process.

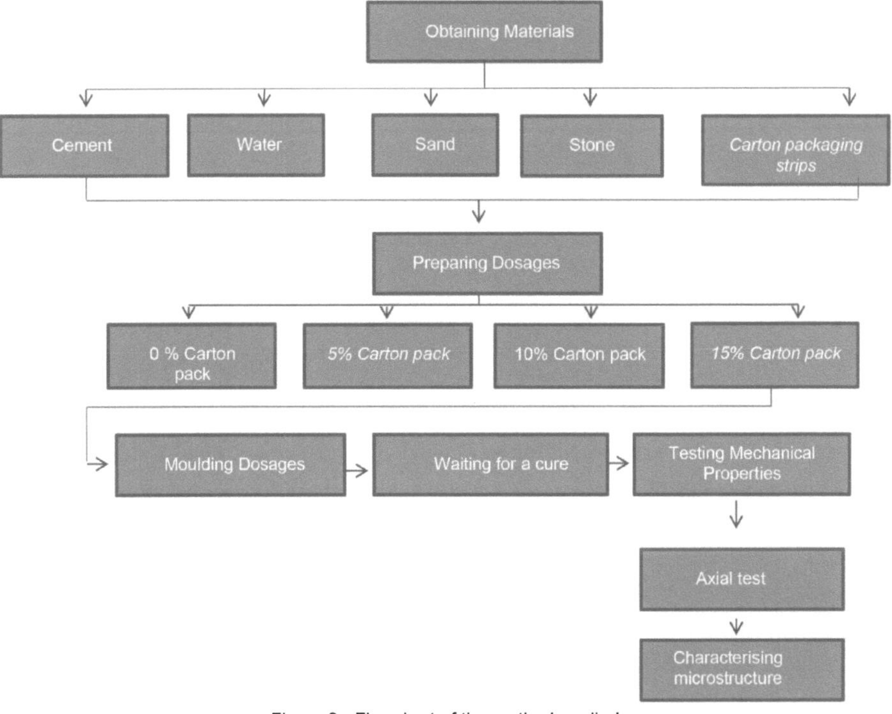

Figure 9 - Flowchart of the method applied
Source: (AUTHOR, 2017)

5.2. 1Material selection

The material chosen for this work was collected from recyclable materials. Thus, 1-litre carton boxes for milk and juice were selected. After separating the cartons, they were sent for removal of the waste and drying, making it possible to preserve the raw material analysed.

The aseptic carton boxes were subjected to two different stages; the first treated the cartons to cut them into 30mm strips, as shown in figure 10. The second stage was guillotine cutting, obtaining 30mm x 3mm strips.

For Schettino (2015), these dimensions were successfully made in his research to cut strips of polyethylene terephthalate (PET).

Figure 10 - Strips of aseptic carton packs
Source: (AUTHOR, 2017)

5.2.2 Preparing the mixtures

In the Engineering laboratory at UniFOA, the materials were used to obtain concrete mixes using strips of cardboard packaging, sand and cement in 1:2:3 mixes. In a ratio of one portion of cement to two of sand and three of gravel, all the material was then poured into the concrete mixer (figure 11) to achieve homogeneity of the concrete.

Figure 11 - Betoneira
Source: (AUTHOR, 2017).

5.2.3 Slump test

In accordance with NBR NM 67-96, the Abrans cone illustrated in figure 12 and the base plate went through an initial humidification process, along with the mould, with water added to fill the Abrans cone, striking the three layers 25 times each, with each layer containing 1/3 of the cone, and then the cone was filled and carefully removed to measure the slump with the opposite cone where we obtained the result of the concrete process at 6cm.

Figure 12 - Abrans cone

Source: (AUTHOR, 2017)

5.2.4 Moulding the specimens

Consequently, after the *slump test,* the moulds are completed with two layers of concrete, adding 12 blows with a socket-shaped rod, as shown in figure 13, around 16 x 60cm in each layer, which must be left for an average of 24 hours.

Figure 13 - Test body socket
Source: (AUTHOR, 2017)

According to NBR 5738, the test specimens are standardised with a cylindrical and diametrical shape of 10X20 cm.

In accordance with NBR 5739-2007, the above-mentioned description subjects the test specimen to filling with a 1:2:3 base mix using 4.5 litres of water and adding 5%, 10% and 15% of carton packaging strips, as described in table 3, and also with the composition without the addition of carton packaging fibres, which serves as a control, as seen in table 2.

Table 2 - Addition sequence of cement, sand, stone and carton packs

Dash	**Cement (kg)**	**Sand (kg)**	**Gravel (kg)**	***Carton packs (%)***	**Water (L)**
T1 (0% strips)	7	14	21	0%	4,5
T2 (5% of strips)	7	14	21	5%	4,5
T3 (10% of strips)	7	14	21	10%	4,5
T4 (15% of strips)	7	14	21	15%	4,5

Source: (AUTHOR, 2017)

After 24 hours, the mould is opened, identifying the specimen with the dosage in each one, and sent to the curing tank, as shown in figure 14.

Figure 14 - CP disinformation
Source: (AUTHOR, 2017)

5.2.5Launch in the wet chamber

The specimens are submerged in the curing tank until the day the compression test is carried out. Every 7 days, three samples will be taken from the tank to carry out the compression test process for each dosage, and measurements will be taken until 28 days have passed since the specimens were moulded, as shown in figure 15.

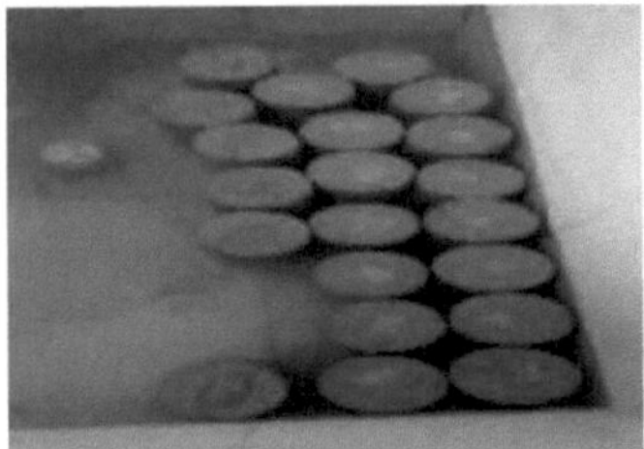

Figure 15 - Launching into the Wet Chamber
Source: (AUTHOR, 2017)

5. 3CHARACTERISATION

5.3.1 Particle size distribution analysis

For this analysis, the aggregates were first weighed on a BEL MARK 5200 fine and coarse aggregate scale. The sieves were then individually introduced into the SOLOTEST sieve shaker in descending order. They were then placed back on the sieve for ten minutes for a new weighing process at a frequency of 15 Hz.

The sieves are designed for two types of aggregate: fine and coarse aggregates. They were selected by decreasing sizes of 2.4mm; 1.2mm, 600pm, 300pm and 150pm, as seen in figure16.

Figure 16 - Sieves for offal
Source: (AUTHOR, 2017)

Coarse Aggregates - The following sieves were selected for this characterisation of the coarse aggregates in descending order: dimensions; 24.0mm, 19.0mm, 12.5mm, 9.5mm, 6.3mm, respectively, as seen in figure 17.

Figure 17 - Sieves: coarse aggregates
Source: (AUTHOR, 2017)

5.3.2 Compressive strength test

The specimens were submitted to the EMIC - PC150C press (shown in figure 18) after being subjected to a curing time of 7 days, starting the compression test of the specimens for capping. Neoprene discs were used to test the compressive strength of the material under investigation. The specimens were tested individually in a hydraulic press machine with continuous loading, without shock and with the presence of tension at a loading rate of (0.05 ± 0.02)MPa/s, with the material breaking after 7, 14, 21 and 28 days.

Figure 18 - Hydraulic press
Source: (AUTHOR, 2017)

5.3.3 SEM analysis

The specimens were morphologically analysed using a HITACHI model TM 3000 SEM, using a secondary electron detector and a tube voltage of 5 kV. The equipment is available in the Materials Characterisation laboratory at UNIFOA Volta Redonda- RJ.

5.3.4 Void index and water absorption

The void ratio and water absorption tests were defined in accordance with NBR 9778/1987:

Step 1 - Drying the samples in an Odontobrás oven at a temperature of 105°C ± 5°C for 72 hours, after which each dry sample was weighed on a Bel model precision balance.

Step 2 - Saturation of the samples in a water tank at a temperature of 23°C ± 2°C for a period of 72 hours, with the sample immersed for the first four hours at 1/3 of its volume. In the following four hours, the sample was immersed at 2/3 of its volume and in the remaining 64 hours, completely immersed, as recommended by NBR 9778/1987. After this period, the samples were weighed.

Step 3 - Weigh the saturated samples at a temperature of 23°C, immersed in water, in order to obtain their immersed weight.

After these procedures, calculations were carried out to determine the water absorption and void index of the samples, which are shown in equations 1 and 2, respectively.

$$A_{Ag} = [(Msat - Ms)/Ms]x100 \qquad (1)$$

Where,

A_{Ag} = Water absorption (%)

Ms = Mass of the oven-dried sample (g)

Msat = Mass of the sample saturated in water at a temperature of 23°C

$$I_V = [(Msat - Ms)/(Msat - Mi)]x100 \qquad (2)$$

Where,

I_v = Emptiness index (%)

Ms = Mass of oven-dried sample (g)

Mi = Mass of the saturated sample, immersed in water (g)

Msat = Mass of the sample saturated in water at 23°C (g)

CHAPTER 6

RESULTS AND DISCUSSIONS

6.1 Particle size distribution

In view of the results presented, the fine aggregate was classified as fine-grained, this classification being characterised by higher water consumption for the satisfactory activity process for the concrete. The fine granulometry test optimised the cement void index, giving the concrete a more compact function with a limited structure, causing a volumetric reduction, as well as restricting the transport spaces for aggressive agents, as shown in table 4 (Coutinho, 2016).

Table 3 - Particle size composition of the fine aggregate

Diameter (mm)	Sieve	Sieve+mould	Withheld	%
2,36	390,80	397,50	6,7	1,34
1,18	333,98	368,14	34,16	6,83
0,60	325,64	611,48	285,84	57,16
0,30	348,03	479,22	131,19	26,23
0,15	342,60	380,77	38,70	7,74
Fund	342,80	347,81	5,01	1,00

Source: (AUTHOR, 2017)

6.2 Compression Resistance

Table 5 shows the compressive strength results of natural cured samples tested after 7, 14, 21 and 28 days of curing, in accordance with NBR 7215 (ABNT 2014). The CPs, with the addition of the carton packaging strips and curing for seven days, showed a considerable oscillation in the test results. The mechanical resistance to compression presented, with the addition of 5%, showed the result shown in Table 5, which was 22.53 MPa for 28 days, which allows non-structural applications for this composition. On the other hand, the 10% and 15% compositions had lower compressive strengths than the 5%, i.e. 20.09 MPa and 16.22 MPa, respectively. Even so, this material can be used in paving, squares, condominium halls and shopping centres, according to NBR 5739 (1994).

Table 4 - Results of Axial Compression Strength Tests on Cylindrical Specimens

Composition	Age	CPS resistance (MPa)	Average (MPa)	Standard Deviation

(0% of strips)	7 days	11,67	9,15	11.32	10,71	1,30
	14 days	16,80	14,11	17.35	16,08	1,70
	21 days	21,85	16,80	19,45	19,36	2,50
	28 days	25,77	23.94	23.88	24,55	1,07
(5% of strips)	7 days	13,39	13,54	14,70	13,82	0,70
	14 days	18,70	18,31	15,35	17,45	1,80
	21 days	20,28	22,15	20,98	21,13	0,94
	28 days	23,48	23,51	20,61	22,53	1,66
(10% of strips)	7 days	9,72	9,22	8,73	9,22	0,49
	14 days	15.61	14,77	14.75	15,04	0,49
	21 days	17,15	18,50	17.92	17.85	0,67
	28 days	20,50	18,98	20,79	20,09	0,97
(15% of strips)	7 days	9,15	8,92	8,48	8,85	0,34
	14 days	9,50	10,27	10,15	9,97	0,41
	21 days	11,74	11,74	8,28	10,58	1,99
	28 days	17,06	18,47	13,13	16,22	2,76

Source: (AUTHOR, 2017)

6.3 SEM

Figures 19 to 21 show the composition of concrete with carton strips, with mixes of 0%, 5% and 15%; and illustrate the SEM images (a) x100 (b) x 500 and (c) x1000 of the fracture surfaces of the concretes after 28 days of curing. In all of them, you can see a microstructure with the presence of carton strips between the pastes and aggregates, porosity and the presence of some microcracks, as shown in Figures 19 (a), (b) and (c). According to Coutinho (2016), the recycled composite increases the adhesion between the aggregate and the paste, due to the physical effect of filling the voids, which makes the matrix denser.

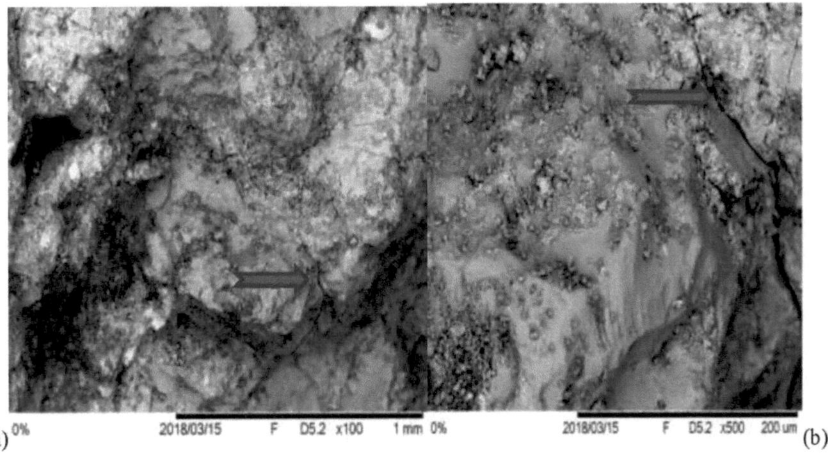

(a) (b)

(c)

Figure 19 - SEM micrograph of natural concrete at x 100 x 500 x 1000 magnification (SEM 28 days)

Source: (AUTHOR, 2018)

Figure 20 shows a view of the magnifications (a) x100 (b) x500 (c) x1000. Micrographs 20(a) and 20(b) show that the finer particles have undergone classic mineralogical formation, microfractures and the interface between the grains agglutinated by the binder. Micrograph 20(c) shows the formation of crystals adhered to the interfaces of the grain boundaries, due to the possible concentration of energy at their edges.

According to Figueiredo 2016, the micrograph shows the altered morphological aspect of the fracture surface due to the greater number of particles present in the composite.

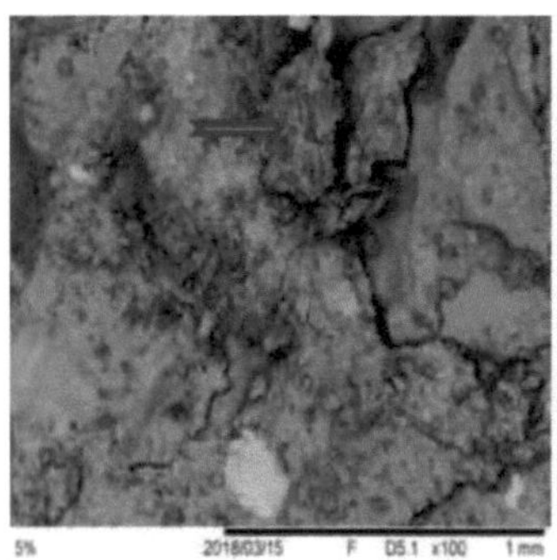

Figure 20 - Micrograph obtained by SEM with 5% action of *carton packaging* strips in concrete at x100, x500 and 1000 magnification (SEM 28 days) **Source:** (AUTHOR, 2017)

The micrograph obtained in the SEM illustrates the morphology with 15% of the carton strip. The micrographs in figure 21 show a view at 100X and 500X magnification of figures 21 (a) and (b) with the presence of the carton aggregate and with cracks in the concrete interface, indicated in the figures, respectively. In figure 21(c) it was possible to identify the existence of the expanded recycled material adhered to the concrete interface as a result of water absorption, a fact that contributes to the loss of the concrete's mechanical strength (Figueiredo, 2016).

Figure 21 - Micrograph obtained by SEM with 15% of the *carton* fibre in the concrete at (a) x 100, (b) x 500 and (c) x 1000 magnification (SEM 28 days)

Source: (AUTHOR, 2017)

6. **4Water** absorption

NBR 9778 - Determinação da absorção de água por Imersão (Determination of water absorption by immersion) was used as a reference to support water absorption by immersion. Table 6 shows the water absorption results.

The average absorption between the specimens with 0% addition was 3.487% and the standard deviation of the samples was 0.090%.

The samples with 5% added *carton* did not show a considerable difference in absorption between the specimens, at 3.481% and a standard deviation of 0.127%.

The samples with 10 per cent added carton strips did not show a considerable difference in absorption between the specimens, at 3.463 per cent and a standard deviation of 0.187 per cent.

The samples with a 15% addition of carton strips did not show a considerable difference in absorption between the test specimens of 3.525% and a standard deviation of 0.007%. Given the results shown in Table 5, it can be said that the average water absorption results were similar between all the compositions. This may be due to the composition of the aseptic carton packs, which have an impermeable characteristic due to the presence of aluminium and polyethylene in their composition.

Table 5 - Water absorption analysis results

Test Body	**CP1**	**CP2**	**Averag e**	**Standard Deviation (I)**
0%	3,423	3,551	3,487	0,090
5%	3,572	3,391	3,481	0,127
10%	3,331	3,596	3,463	0,187
15%	3,530	3,520	3,525	0,007

Source: (AUTHOR, 2017)

6.5 **Emptiness** ratios

When comparing the void ratios with the SEM micrographs, the images are consistent with the values in Table 6. The void ratio on the surface of the material showed an average of 1.268% and a standard deviation of 0.028% for the composition without addition.

When comparing the Cp with the addition of 15% recyclable material to the SEM, the representation of the micrographs associated with the numerical results, the index of voids on the surface of the concrete was visualised, with an average of 1.834% and a standard deviation of 0.624%. The results are similar between the compositions, which is due to the low interaction of the carton fibres at the interface with the concrete (López, 2016).

Table 6- Result of the void indices

Test Body	CP1	CP2	Average (%)	Standard Deviation (%)
0%	1,248	1,288	1,268	0,028
5%	1,760	1,719	1,739	0,029
10%	1,580	1,583	1,581	0,002
15%	2,276	1,393	1,834	0,624

CHAPTER 7

CONCLUSIONS

The *carton packaging* strips show specific characteristics for reuse in the production of concrete for the construction industry, presenting themselves as a sustainable alternative, rich in recyclable raw materials, positively optimising the concept of preserving the ecosystem, eliminating the accumulation of unused waste in the country.

In the comparative tangent of the compressive strength tests on concrete specimens, it was noted that with the addition of *cartonboard, there was* a reduction in strength: 22.53 MPa for 5% addition of *cartonboard, 20.09MPa* with 10% addition and 13.13MPa with *15% addition of cartonboard* strips. The micrographs show that at 28 days there is no good interaction between the concrete and the carton strips.

The product produced has non-structural applications, such as paving streets with little traffic and squares.

Based on the results of the water absorption tests, the results for natural aggregate had an average of 3.487 MPa and a standard deviation of 0.090 MPa. According to NBR 9781:2013, absorption of up to 6% is permitted for natural aggregate.

The average water absorption results were higher in the composition with 15% recycled material added, with an average of 3.525% and a standard deviation of 0.007%, which was to be expected due to the higher percentage of carton packs.

The percentages obtained for water absorption and void ratio were decisive for the conclusion of this book. The void ratio calculations showed that the natural concretes, without the addition of recycled material, showed a reduction in interface voids, with results averaging 1.268% and a standard deviation of 0.028.

Cps with 15% added carton strips showed an increase in surface voids, a phenomenon clearly observed with the help of SEM. The results were 1.834% for the mean and 0.624% for the standard deviation.

CHAPTER 8

SUGGESTIONS FOR FUTURE WORK

As it is impossible to cover all the studies on the behaviour of a material in a single study, suggestions are made for future work.

Waterproof the carton strips

M Micronise the size of the aseptic carton fibres, which will be added to the concrete

C Making concrete floors with the addition of aseptic carton packs

REFERENCES

ABNT Associação Brasileira De Normas Técnicas NBR 5733:1991 High early strength Portland cement - Specification. Rio de Janeiro

ABNT. Brazilian Association of Technical Standards (2006). *NBR. 5738- Moulding and curing of cylindrical or prismatic concrete specimens.* Rio de Janeiro.

ABNT. Brazilian Association of Technical Standards. NBR 5.732- Ordinary Portland cement - specification. Rio de Janeiro, 1991. 5p.

ABNT. Brazilian Association of Technical Standards. NBR 5739 - Concrete - Compression test of cylindrical specimens, 2007.

ABNT. Brazilian Association of Technical Standards. *NBR* 6118: Design of Concrete Structures." *Rio de Janeiro* (2014).

ABNT. Brazilian Association of Technical Standards. NBR 7215. Determination of the compressive strength of Portland cement. Rio de Janeiro (2008).

ABNT. Brazilian Association of Technical Standards. *NBR. 9778- Hardened mortar and concrete - Determination of water absorption by immersion - Void index and specific mass.* Rio de Janeiro (2005).

ABNT. Brazilian Association of Technical Standards. NRB NM 67: Concrete: - Determination of consistency by cone trunk slump. Rio de Janeiro:

AJAM, H. K. (2013). Utilisation of sheredded tetra-pak in hot mix asphalt. *Al-Qadisiyah Journal for Engineering Sciences,* **6**(3), 287-293.

ALVES L.F.S.J. Composites obtained by incorporating shredded long-life packaging waste into concrete roof tiles: Preparation and mechanical characterisation. ENEPEX- Dourados-

MS /2013.

Amaral, P. T. D. (2017). ª*Analysis of 2nd order effects in compressed reinforced concrete bars* (Doctoral dissertation, University of São Paulo).

AMBROZEWICZ, Paulo Henrique Laporte. **Construction materials**. 2012.

Brazilian Association of Technical Standards. NBR 7.214. Standard for cement testing. Rio de Janeiro, 1982. 7p.

BAUER, L. A. **Construction materials.** 5ed. Rio de Janeiro: LTC, 2011,471 p.

BENTUR, A.; MINDESS, S. Fibre Reinforced Cementations Composites. London and New York: Modern Concrete Technology Series, 601 p. 2 ed. 2007.

BETSUYAKU, Renato Yochio. **Construction of eco-bricks with the addition of diatomaceous sand**. 2015. Dissertation (Professional Master's Degree in Materials) - Volta Redonda University Centre.

BRAZIL. National Solid Waste Policy (Law No. 12.305/2010). Brasília: Diário Federal Official Gazette, 2010. Available at . Accessed on 20.05.18.

CALLISTER jr; WILLIAN, D. **Materials Science and Engineering**. An introduction. 9ª Ed.LTC- Livros Técnicos e científicos S.A: Rio de Janeiro, 2016.

CARASEK, H. **Construction materials and principles of materials science and engineering**. Ed. G.C. Isaias- São Paulo: IBRACON, 2007.

CASTANHEIRA, Bruno César Angelini. Study of the quality of stabilised mortars used in structural masonry. 2017.

CASTRO, A.L. Fibres and polypropylene and their influence on the behaviour of concrete exposed to high temperatures. Revista Cerâmica 57,2011 pag. 22-31.

CEMPRE, 2014. National Solid Waste Policy - The impact of the new law against global warming, Business Commitment for Recycling (CEMPRE) and the Packaging Technology Centre of the Food Technology Institute (CETEA/ITAL).

CONAMA. National Environment Council. **Resolution No. 258 of 26 August 1999.** Brasília 1999.

COUTINHO, Rogério de Oliveira. Production of paver from concrete composite with reuse of solid waste generated in civil construction. Volta Redonda: UniFOA, 2016.

CUNHA, C. Salles. **Concrete mixes without structural function with the addition of rubble**. 2015. Dissertation (Professional Master's Degree in Materials) - Volta Redonda University Centre

DIAS, R. Comparison of the determination of the toughness of steel fibre reinforced concrete using the wedge opening test and ASTM C1609. 2018.

FATHI, A., SALIH, M., SHAFIG, N., NURRUDIN, M.F., ELHEBER, A. and MEMON, F.A. (2014), **"Comparison of the 150 Experimental investigation on** self-compacting concrete reinforced with steel fibers effects of different fibers on the properties of self-compacting **concrete", Res. J. Appl. Sei. Eng.** Tech., 7(16), 3332-3341

FIGUEIREDO, L.R.F **Ultra high molecular weight polyethylene (UHMWPE) and quasicrystal (AlcuFe) composite: Thermal and mechanical behaviour.** Federal University of Paraíba - 2013

FIORELLI, J. Evaluation of the thermal efficiency of recycled tiles made from long-life packaging, 2008.

HWANG, J.-H.; LEE, D.H.; JU, H.; KIM, K.S.; KANG, T.H.-K.; PAN, Z.: Shear deformation of steel fibre-reinforced prestressed concrete beams. Int. J. Concr. Struct. Mater. 10(3), 53-63 (2016)

KHALOO, A., Molaei Raisi, E., Hosseini, P. and Tahsiri, H. **(2014), "Mechanical performance of selfcompacting concrete reinforced with steel fibres", Constr.** Build. Mater., 51, 179-186.

LACERDA, C. **Mortar traces using electric steel slag.** 2015. Dissertation (Professional Master's Degree in Materials) - Volta Redonda University Centre.

LÓPEZ,M.M."Physicochemical modification by gamma radiation of recycling materials from tetra pak containers and their reuse in polymer concrete.2016 Postgraduate Programme in Environmental Sciences at the Faculty of Chemistry of the Autonomous University of the State of Mexico.

METHA, P. K. and MONTEIRO, P.J.M. Concreto: Estruturas, Propriedades e Materiais. Editora Pini, São Paulo, (2010).

MONTE, R. Characterisation and control of the mechanical behaviour of fibre-reinforced concrete for pipes. 2015. 158 f. Thesis (Doctorate) - University of São Paulo, São Paulo, 2015.

MOREIRA, G.C. Alternative aggregates in the mode-mould fracture of polymeric mortars, 2017.

NAAMAN, A. E. Engineered Steel fibres with Optimal Properties for Reinforcement of Cement Composites. Journal of Advanced Concrete Technology. Japan Concrete Institute, v. 1, n. 9, p. 241-252, 2003.

NEVES, A. C. R. R.; CASTRO, L. O. A. **Separation of recyclable materials: panorama in Brazil and incentives to practice**. Electronic Journal on Environmental Management, Education and Technology, v.8, p.1734-1742, 2012.

NEVILLE, A. M. Properties of Concrete. 5.ed. Porto Alegre: BOOKMAN, 2016.

NOSHEEN, H. An Investigation on Shear Behaviour of Prestressed Concrete Beams Cast by Fiber Reinforced Concrete. *Arabian Journal for Science and Engineering,* 2018, 1-9.

PATIL, S.P.M.; PAWAR, M.M.: Prediction of shear strength of steel fibre reinforced concrete beams without web reinforcement. Int. J. Eng. Res. Technol. **4**(04) (2015)

PEDROSO, M. C. **Sustainability in the reverse supply chain: a case study of the Plasma Project**. Revista de Administração, v.42, p.414-430, 2007.

PEREIRA. R.D.A. Um estudo dos canais reversos em uma empresa de embalagens cartonadas/ Rio de Janeiro XXVIII Encontro Nacional de Engenharia de Produção 2008.

SALVADOR, R.P "Comparative evaluation of the mechanical behaviour of synthetic and
steel fibre-reinforced concrete Revista Matéria, v.18,n.2,pg.1273-1285,2013.

SCHETTINO,R.M. **Concrete with the addition of ethylene polyteraphthalate strips**. 2015. Dissertation (Professional Master's Degree in Materials) - Volta Redonda University Centre.

SILVA,K.C.P **Reuse of Tetra Pak packaging waste in roofing** .Bras. Eng. Agríc. Ambiental, v.19, n.1, p.58-63, 2015.

TORRES, D.R.**Influence of variation in rotation speed and type of cement on the properties of coating mortars in the fresh and hardened states.** Revista Cerâmica, n.368 dec.2017.

UEMURA, M.R.B. Reverse logistics of carton packs and the reduction of greenhouse gas emissions: the Tetra Pak case.SEMEAD, 2017 pg de 1-13.

WU, X., KANG, T. H. K., LIN, Y., & HWANG, H. J. (2018). Shear Strength of Reinforced Concrete Beams with Precast High-Performance Fiber-Reinforced Cementitious Composite Permanent Form. *Composite Structures.*

Printed by Books on Demand GmbH, Norderstedt / Germany